Basic Tips

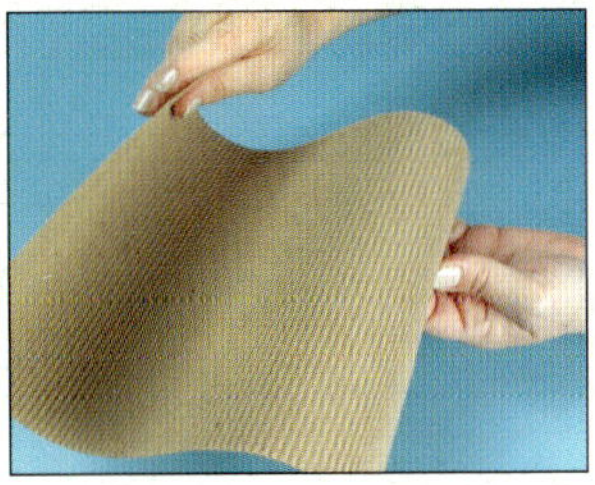

Bend or Roll corrugated paper with ridges going across the roll.

Score Papers. Draw a heavy line with a pen on back side of paper. This will crush the fibers enough to make the paper easy to bend.

Stack Paper. Change the direction of the corrugation with each layer to make a sturdier stack.

Color Paper. Rub a multicolor ink pad over paper. The ridges, ribs and bumps will pick up ink leaving recessed areas untouched.

Adhesives. E6000 or Goop - Use to attach hardware, legs and ornaments.
Thick Tacky Glue - Use for layering, wrapping and coiling paper.
Sticky Dots - Use to replace tacky glue.
Double Sided Tape - Use for layering.

Table of Contents

NOTE

• **What is Corrugated Cardboard?** Corrugated cardboard consists of two flat pieces of cardboard with a wavy piece of cardboard sandwiched between them. This makes a stiff sturdy surface.

• **What is Corrugated Paper?** Corrugated paper consists of one flat piece of paper with a corrugated piece of paper on top. The paper can be found in many colors including Natural. The pattern of the corrugation can be straight up and down or wavy as found in the Wavy Ridges corrugated paper.

you're
a
star
House
House

General Card Instructions

MATERIALS:
5½" x 4¼" piece of Natural Kraft DMD Corrugated cardboard • 5½" x 8½" piece of desired color cardstock • Gold extra fine point marker • Black ultra fine point permanent marker • Punches (giant flower, house, sun, maple leaf, paw, heart) • Double-sided adhesive tape • Tacky glue

INSTRUCTIONS:
Fold cardstock in half to form a 4¼" x 5½" card. Place 3 strips of double-sided adhesive on back of corrugated paper, one down each side and one down center of card back. Remove paper backing and attach corrugated paper to front of card. Punch design through corrugated and cardstock layers on front of card. Add message.

Black Paw Card - Using Black cardstock, punch paw in the lower left corner 1" from bottom of card. Write 'DOG' with Gold marker in cut out paw. Write message "Here's hoping you have a - - - gone great birthday!" inside card, fitting around word 'dog'. Glue paw punch on right inside section 1" down from fold.

Green Maple Leaf Card - Using Dark Green cardstock, punch a maple leaf 1" from bottom of card and 1" in from right side. Write 'THANKS' with Gold marker in cut out. Write message "- - - so much!" inside card, under 'Thanks'.

Yellow Star Card - Using Yellow cardstock, punch a star 1" from bottom and 1" from right side of card. Write 'STAR' in cut out with Black marker. Write message "Congratulations you're a - - - !" inside card, fitting around 'Star'. Glue the star punch out on lower left corner inside card.

Peach Flower Card - Using Peach cardstock, punch a Giant Flower in front of card. Write 'FLOWER' in cut out using Black marker. Write message "A friend is a - - - - - - in the garden of life!" inside card, fitting around 'flower'. Punch a ¼" hole in center of flower and glue to inside of card on bottom left corner.

Burgundy House Card - Using Burgundy cardstock, punch 2 houses on front of card. Push punch as far up as possible on bottom of card front and punch houses 1" from sides of card. Write 'HOUSE' with Gold marker in each cut out. Write message "Greetings from our - - - - - to your House!" inside card, fitting around 'House'.

Burgundy Heart Card - Using Burgundy cardstock, punch a heart in lower right corner. Write 'LOVE' using Gold marker in cut out. Write message "Sending you all my - - - -" inside card with Gold marker, fitting it around 'Love'. Glue heart punch out in upper right corner inside card.

Sunshine Card - Using Terra Cotta cardstock, punch Sun 1" from bottom of card and 1½" from right side. Write 'SUN' using Gold marker inside cut out. Write message "You are my - - - shine!" inside card, fitting around 'Sun'.

Stationery Box & Notecards

by Vivian Peritts

Card Holder

MATERIALS:
12" x 16" sheet of wavy Ridge corrugated paper • 6" x 6" sheet of Burgundy corrugated paper • 3" x 3" sheet of Natural Kraft corrugated paper • 1½ yards of sisal • Tacky glue • Pinking scissors • ¼" and 1¼" circle punches • Beads

INSTRUCTIONS:
Cut out card box. Place a straight edge along fold line and fold up on all fold lines. Glue bottom of box up. For top closure, cut the following pieces with pinking scissors: (A) Burgundy corrugated paper, (B) Wavy Ridges and (C) Natural Kraft corrugated paper. Punch two 1¼" Burgundy corrugated circles. Punch holes side-by-side in all 5 pieces. Glue pieces together aligning holes. Cut a 10" piece of sisal. Fold sisal in half and push folded end through 15mm Brown bead. Bring loop back to top of bead and push one of the free ends through loop. Loop is between cords. Push sisal ends through 20mm Brown bead and 40 x 10mm Green barrel bead. Push one end of sisal though each of the holes in the closure. Tie a knot with sisal ends on back. Glue closure to flap of box. Cut the bottom closure from wavy ridges with ridges going up and down. Punch holes in closure. Cut remaining sisal into 3 equal pieces and tie all 3 pieces in an overhand knot near one end. Braid sisal for 2". Push ends into second hole and tie an overhand knot. Glue closure to bottom of box ¼" down from top of bottom section. Let dry. Fasten box by closing lid and pushing beads through loop.

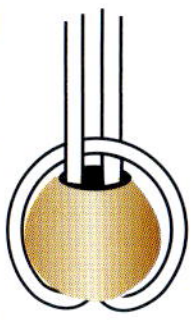

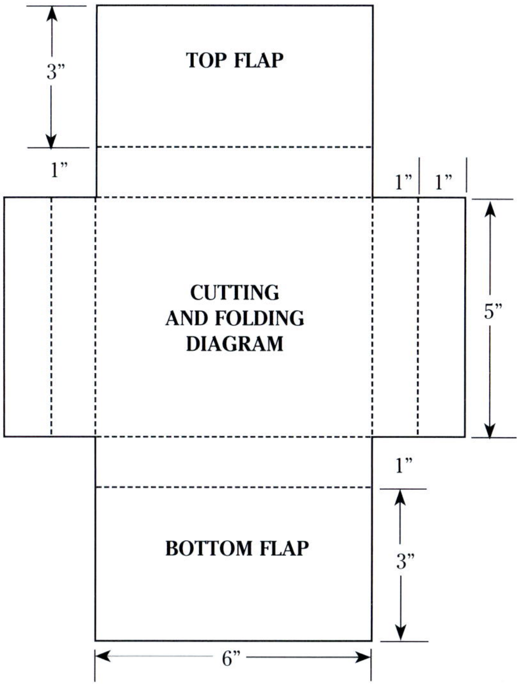

Fold cardstock in half, attach corrugated paper and punch design through front of card.

Picture Frames

by Vivian Peritts

Rectangular Frame

MATERIALS:
Small rectangular frame • 8½" x 11" sheets of corrugated paper (1 sheet each Burgundy and Hunter Green, two sheets Natural kraft) • ¾" x 18" strip of Natural wavy ridges paper • Pinking scissors • Jumbo heart punch • Tacky Glue

INSTRUCTIONS:
1. Trace around inside and outside of frame on the back of Burgundy paper. Cut out and glue on frame.
2. Cut wavy ridges paper into ⅝" strips. Glue around outside of frame.
3. Punch 12 Jumbo hearts from Natural kraft paper. Arrange in 4 stacks of 3 hearts. Glue stacked hearts together and glue one stack in each corner with points of hearts pointing to corners.
4. Refer to diagram for size and colors of corrugated paper in triangles.

Oval Marigold Frame

MATERIALS:
4½" x 4" cardboard oval frame • ⅝" x 15" strip of Black wavy ridges corrugated paper • 6" x 6" sheet of Marigold Corrugated paper • Thirty-three 1¾" wood matches • Tacky glue • Mat knife

INSTRUCTIONS:
Trace around frame inside and outside on smooth side of Marigold paper. Cut out with scissors and mat knife. Glue on front of frame. Glue Black wavy ridges around outside of frame. Light matches. Burn heads only and blow out. Glue on frame with the heads facing in.

Glue Black wavy ridges around outside edge of frame.

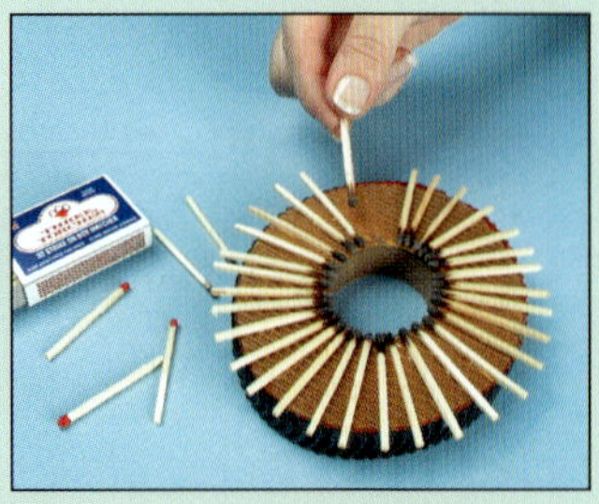

Glue matches evenly spaced around front of frame.

Square Frame

MATERIALS:

Natural cardboard 9" x 9" x 5/8" frame • Sixty-six 1¾" wood matches • 8½" x 11" sheets of corrugated paper (1 each White, Hunter Green and Natural kraft) • 2" x 18" strip of Black wavy ridges corrugated paper • Wavy edge scissors • Tacky glue

INSTRUCTIONS:

1. Cut Black wavy ridges paper into ½" strips and glue around inside and outside edges of frame.
2. Use scissors to cut out 3 square mats for top of frame as follows: 6½" square White with a 4½" square opening, 7½" square Natural with a 4" square opening and 8" square Hunter Green with 3" square opening. Glue on frame starting with largest mat.
3. Light matches and burn heads only. Blow out. Glue matches to frame referring to photo and leaving ½" between matches and alternating heads. The 3 matches at each corner have heads pointing to corners.

Make an easy frame for every photo or print.

Little Trunk

by Vivian Peritts

MATERIALS:
Unfinished 11" x 3" x 4" hinged wood candle box • Two ½" unfinished drawer pulls • Four ½" flat buttons • Four 1" x ⅜" toy wheels • Four 1 3/16" x ⅞" spools • One 4 prong dinner fork • One latch hook with eye screw • Wood n Resin Cherry Gel Stain • 12" x 6" Copper Metallic Corrugated paper • 12" x 16" sheet of Natural Wavy Ridges • 8½" x 11" Corrugated Paper (2 sheets each Hunter Green and Burgundy) • Needle nose pliers • Thick tacky glue • E6000 glue • Pinking scissors and Rockies Canyon Cutters

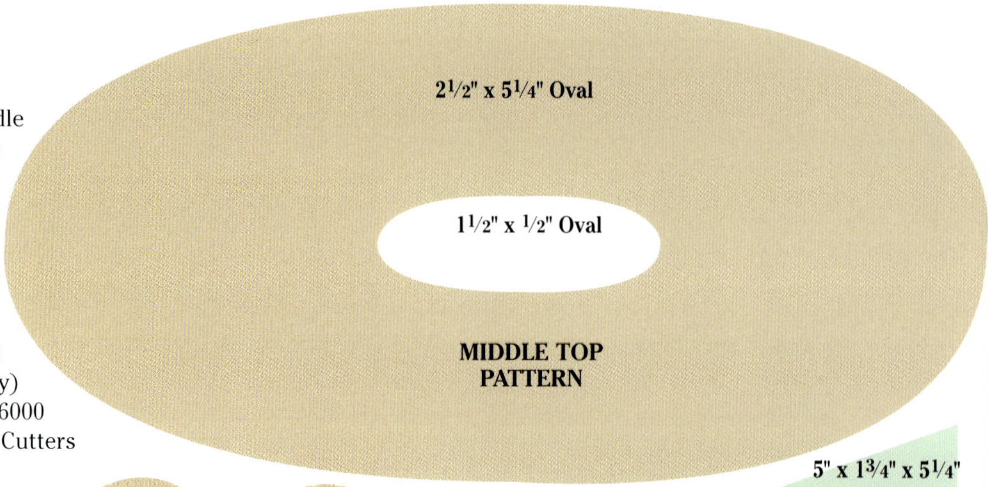

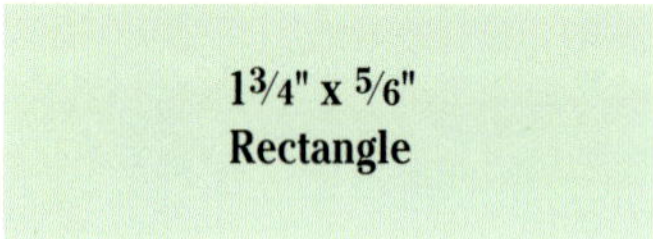

INSTRUCTIONS:
Prepare wood. Paint box with wood stain and wipe off excess with a soft cloth. Let dry. Using Rockies scissors, cut a 10¼ x 3¾" Copper corrugated rectangle. Glue on top of box.

Center Design on Box Top - Using pinking scissors, cut a 1½" x ½" oval of Natural wavy ridges. Glue on Green corrugated paper and using pinking scissors, trim slightly larger than Natural piece. Glue on top of a sheet of Burgundy paper and repeat cutting and gluing through layers, trimming each layer slightly larger than previous piece adding the following layers: Natural wavy ridges, Green, Burgundy, Natural wavy ridges, Green, Burgundy, Natural wavy ridges and Green. Using pinking scissors, cut a 2½" x 5¼" Natural wavy ridges oval and trim a Green oval slightly larger. Glue large oval combination on center top of box. Glue small oval on large oval.

Fork Accent - Use needle nose pliers to curl fork tines. Curl one tine tight and curl each remaining tine slightly less. Bend third tine down. Bend handle up 1" from wrist of fork, 1" further up bend handle down and 1" from last bend, bend the remainder of the handle parallel to the bottom of the fork. Mount fork on top of center oval with E6000.

Sides of Box Top - Make 2. Cut with pinking scissors, glue and stack. Cut each new layer slightly larger than previous layer. Start with a ¾" pinked Natural wavy ridges circle and add the following layers: Green, Burgundy, Natural wavy ridges, Green, Burgundy, Natural wavy ridges and Green. Cut a ¼" Natural wavy ridges circle with pinking scissors and mount it on Green paper. Trim Green paper slightly larger than the previous layer with pinking scissors. Glue one of the large circle combinations on each end of the top and glue smaller circle combinations on larger circles. Glue a wooden drawer pull on top of each stack.

Edges of Lid - Cut Green corrugated paper to match all 4 sides of lid. Trim each piece with pinking scissors and glue on lid.

Box Bottom Front Sections - Cut wavy ridges paper to match lower front, lower back and both lower sides of box. Trim each piece with pinking scissors and glue on box. Cut 2 Green triangles with pinking scissors. Glue triangles on lower front of box. Cut 2 Natural wavy ridges triangles as shown. Glue Natural wavy ridges triangles to Burgundy paper and trim last layer slightly larger. Mount on Green triangles. Top with 1" Burgundy pinked circles on slightly larger Copper circles. Add wood buttons and glue to corners of the front.

Latch - Attach latch with eye screws to center of front of box lid. Attach eye hook to lower box front.

Ends of the Box - Make 2 stacks. Cut a Green 1¾" x ⅝" rectangle with pinking scissors. Mount rectangle on Burgundy paper and trim paper slightly larger with pinking scissors. Repeat gluing, stacking, and trimming through next layers: Natural wavy ridges, Green, Natural wavy ridges and Burgundy. Mount on Natural wavy ridges covered ends of box. Top each stack with a button.

Legs - Wrap centers of 4 spools with Green paper. Glue each spool on top of a wheel and use E6000 to glue spools to bottom of box.

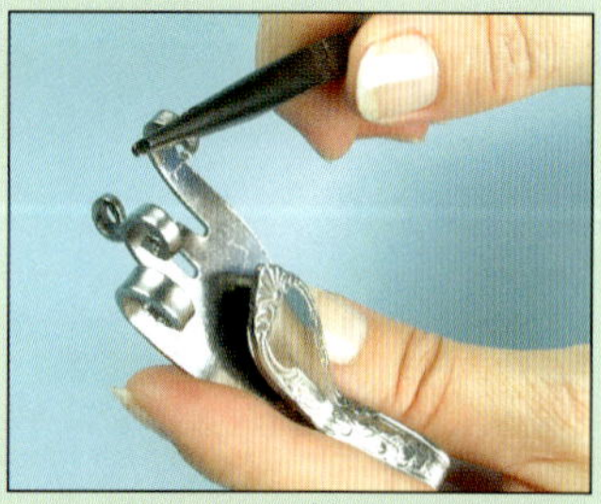

Curl fork tines using needle-nose pliers. Shape handle.

Attach latch to center front of box.

Wrap centers of spools with wavy paper, glue in place.

Cut papers with decorative scissors and punch designs.

Stack and glue paper layers together.

Cut paper strips and roll to make coils.

Cut tapered strips. Roll and glue to make legs.

Treasure Box

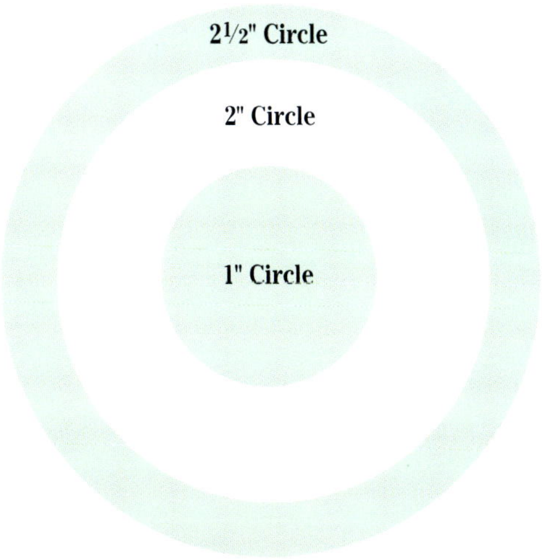

by Vivian Peritts

MATERIALS:
Mixed Nuts nut case • Corrugated Paper (12" x 12" sheets - 2 Walnut and 1 Navy, $8\frac{1}{2}$" x 11" sheets - 4 Navy, 3 Marigold, 4 Burgundy and 1 Ivory) • 8" x 12" sheet of Silver Metallic Corrugated paper • One 12" x 16" sheet of Natural Mini-Flute Corrugated paper • One 12" x 16" sheet of Natural wavy ridges paper • Five $1\frac{3}{8}$" x $\frac{3}{8}$" wood wheels • Five 1" x $\frac{3}{8}$" wood hearts • Six sheets of Sticky Dots Adhesive • Tacky glue • Pinking and Scallop scissors, Rockies and Pompeii Canyon Cutters • Punches (giant heart, $1\frac{1}{4}$" circle, long reach $\frac{5}{8}$" circle, $\frac{1}{4}$" hole, $\frac{1}{4}$" square) • One $2\frac{1}{2}$" Silver nail

INSTRUCTIONS:
1. Assemble box following the package instructions.
2. Cut walnut paper as follows: front of box - $11\frac{1}{2}$" x $2\frac{3}{4}$", top of lid - $11\frac{1}{2}$" x $8\frac{1}{4}$", top back - $11\frac{1}{2}$" x $2\frac{1}{8}$". Trim with pinking scissors and apply to box with sticky dots. Cut 2 box sides of Walnut paper to match sides and apply.
3. Cut 18" x $\frac{5}{8}$" strips of Wavy Ridges Natural Paper and cover sides of box along edge and top edges of back piece. Cut 18" x $\frac{1}{2}$" strip of Wavy Ridges and apply to front of box lid.
4. Cut 8" x $10\frac{1}{2}$" piece of Navy paper and trim with pinking scissors. Apply to top of lid. Cut 8" x $10\frac{1}{2}$" Natural Corrugated with Pompeii scissors, punch a $\frac{1}{4}$" circle in each large loop and apply to lid.
5. Cut $9\frac{1}{2}$" x $6\frac{1}{2}$" piece of Ivory Corrugated paper and apply to lid. Cut $9\frac{1}{2}$" x $6\frac{1}{2}$" piece of Burgundy corrugated paper with Rockies scissors. Punch a $\frac{1}{4}$" square in each outward point. Apply to lid.
6. Cut $7\frac{1}{2}$" x 5" piece of Natural Wavy Ridges paper with straight edge scissors and apply to lid. Cut 7" x $4\frac{1}{4}$" piece of Silver corrugated paper with pinking scissors and apply to lid.
7. Cut $6\frac{1}{4}$" x 4" piece of Navy corrugated paper with Rockies scissors. Punch $\frac{1}{4}$" square in each outward peak. Apply to lid.
8. Cut and apply remaining pieces. Make each slightly smaller than the previous one. Cut $5\frac{1}{4}$" x $2\frac{1}{2}$" piece of Marigold paper. Cut Burgundy, Natural, Navy, Marigold and Burgundy. Trim each piece with pinking scissors.
9. Top lid with 1" scalloped Natural Wavy Ridges circle on a 1" Burgundy circle.
10. Cut 2 pieces Natural Wavy Ridges paper using pattern and apply to sides of box on curved section. Cut 2 Natural Wavy Ridges 5" x 4" x 6" triangles. Apply to back bottom corners of each side. Trim triangles with pinking scissors progressively smaller in the following colors: Burgundy, Silver, Navy and Natural Wavy Ridges.
11. Cut $2\frac{1}{2}$" Burgundy corrugated circles with pinking scissors and apply next to side of box shape cut with pattern. Glue a $1\frac{3}{4}$" x $1\frac{3}{4}$" x $1\frac{3}{4}$" Natural Wavy Ridges triangle below circle.
12. Cut a 2" Marigold square, apply to side front bottom corner. Top with a pinked Navy square and Wavy Ridges slightly smaller pinked squares. Above Marigold square, glue a 2" x $1\frac{1}{8}$" pinked Wavy Ridges rectangle.
13. Cover wheel tops and sides with Marigold. Glue one Marigold wheel on top of Burgundy circle and a large Navy coil on top of front side squares.
14. For top back and front of the box, cut strips of Natural Mini-flute corrugated paper with decorative edge scissors referring to photo. Apply to box.
15. To the bottom front of the box, add a Marigold wheel to center. Cover hearts with Burgundy. Add a heart on each side.
16. For lid handle, cut two 11" x $1\frac{1}{2}$" strips of Navy corrugated paper. Place end to end. Mark one end $\frac{3}{4}$" wide. Cut at an angle to make gradually shorter. Start at the $1\frac{1}{2}$" end and wind strip into an oval. Keep winding and add a second strip and stop at the $\frac{3}{4}$" end. Keep top of roll even. Cut $\frac{3}{4}$" strip of Burgundy corrugated strip and wind around once. Push the nail through center of knob and push nail into front of lid, glue.
17. For back top edge of box, cut five $\frac{1}{2}$" Navy strips. On first strip, roll ends in $2\frac{1}{2}$", glue the center 6" to top of box. Roll second and third strips the same way, but 2" in from one end and 4" in from other end leave 2" unrolled and glue on sides of first strip with small coils to the outside. Glue 2 Marigold wheels in center. At the ends make 2 small Navy coils. Punch 2 Natural $\frac{5}{8}$" circles and glue one on front of each coil. Glue to ends. Stack and glue wheels and hearts as shown.
18. For each leg, cut 2" x 11" strip of Navy corrugated paper. Mark one end $1\frac{1}{2}$" from bottom corner and draw tapered line to 2" end. Cut along marked line with pinking scissors. Cut $1\frac{1}{2}$" x 11" strip of Navy corrugated paper. Mark one end $\frac{3}{4}$" wide and draw tapered line to $1\frac{1}{2}$" end. Cut strip with pinking scissors on line. Roll first strips starting at the 2" end. Glue as you roll. Attach $1\frac{1}{2}$" end of second strip at end of first strip and continue gluing and rolling strips. Secure with rubber bands until dry. Make 4 legs. Glue on bottom of box.

Miniature Desk

by Vivian Peritts

MATERIALS:
Unfinished $10\frac{1}{8}$" x $14\frac{1}{4}$" x $7\frac{1}{8}$" Dolly Desk with hinged lid • Four Silvertone bun shaped drawer pulls approximately $1\frac{1}{4}$" diameter • Four $1\frac{1}{2}$" nails • One Zinc $\frac{5}{32}$" x 2" cotter pin • Silver 14 x 9mm oval bead with large hole • One dinner fork • One cocktail fork • One Silver eye screw and 12" of Silver chain • Two $1\frac{3}{8}$" x $\frac{3}{8}$" wood wheels • Needle nose pliers • Two U shape staples • Hammer • Pinking scissors • $1\frac{1}{4}$" circle punch • Hunter Green Heavenly Hues paint • Clear Gel Stain • $8\frac{1}{2}$" x 11" sheets of Corrugated Paper (3 sheets each of Hunter Green, Burgundy, Off White and Black) • E6000 glue • Tacky glue

INSTRUCTIONS:
Prepare wood. Mix $\frac{1}{3}$ Green Heavenly Hues with $\frac{2}{3}$ Clear Gel Stain and paint box. Wipe any excess paint off with a soft rag. Let dry.

Top of Desk - Referring to patterns for top of desk, cut out 2 sets of designs, one of pattern and one reverse of pattern. Trim each piece with pinking scissors and glue on lid of desk.

Front of Desk - Cut out 2 sets of triangles for front design. Reverse the design for each side, as shown. Begin with Green corrugated triangle 2" wide x $\frac{5}{8}$" high. Trim with pinking scissors, glue to Off White, trim larger, glue to Burgundy, trim larger, glue to Hunter Green, trim larger, glue to Off White, trim larger and glue all on front of desk.

Side of Desk - Cut out 2 sets of side design, reversing one set. Trim with pinking scissors and glue on sides of desk. Cover wheels with Burgundy and add to top of design.

Upper Back of Desk - Cut out all pieces for upper back. Cover standing portion of back with Black corrugated paper. Trim large Black piece with pinking scissors and glue on back of desk. Using pinking scissors, cut triangles beginning with Black corrugated paper $5\frac{1}{4}$" x $\frac{7}{8}$ triangle. Glue onto Off White, trim larger, glue to Hunter Green, trim larger, glue to Burgundy, trim larger, glue to Off White, trim larger. Glue all to Black corrugated paper on back. Cut a 11" x $\frac{1}{4}$" strip of Burgundy paper with corrugation running short way. Curl loosely from each end. Cut a 11" x $\frac{1}{2}$" strip of Green corrugated paper with the corrugation running short way. Coil tightly and glue. Glue both pieces over triangles as shown.

Top of the Upper Back of Desk - Cut three 11" x $\frac{1}{4}$" Green and one 8" x $\frac{1}{4}$" Burgundy strips with corrugation going in short direction. Coil one Green strip from both ends leaving 3" uncoiled in center. Glue on top of desk back with coils turned wide. Coil Burgundy strip from both ends leaving 1" straight in center. Glue strip on top of Green strip. Coil last 2 strips into one large coil and glue on center top of desk back.

Dowel Rod Covers - Three dowel rods connect the desk's legs on sides and the back. For back dowel, cut one Off White and one Burgundy triangle. Trim Burgundy triangle along sides with pinking scissors. Glue around center of dowel with Burgundy on top of Off White triangle. To cover both sides of dowels, cut 2 sets of Off White and Green triangles. Pink sides of Green triangle. Glue around the sides of dowels with Green on top of White.

Legs - Cover front and outside only of each leg. Cut eight $\frac{1}{2}$" wide Green strips the length of the leg part. Piece strips if necessary. Corrugation should run crosswise on the strips. Trim each strip with pinking scissors and glue on legs. Cut 8 strips of Burgundy with corrugation running up and down, the width of 2 corrugation ridges and long enough for the legs, glue on Green strips.

Feet - Hammer one nail in center bottom of each leg so that knob can fit over nail and be flush with the bottom of leg. Glue in place using E6000.

Closure - Use needle nose pliers to bend forks. Use E6000 and U staple to secure forks. Attach chain to eye of cotter pin. Glue bead on cotter pin close to eye. Attach chain to eye screw and screw into bottom of desk.

Note: When attaching forks, make sure 3 curled tines of cocktail fork match 2 lower curled tines of dinner fork. The cotter pin should fit through all 5 tines to secure lid.

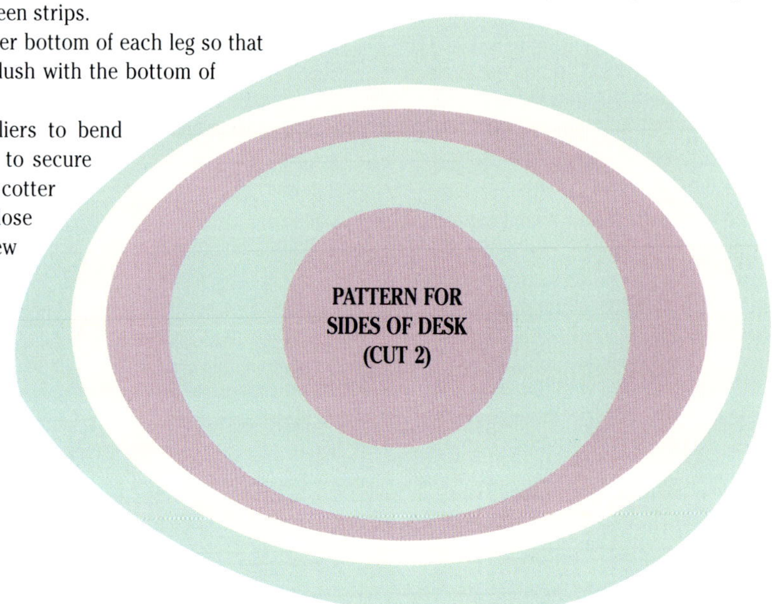

Cut triangles of paper, wrap around centers of dowels.

Bend fork tines. Attach with glue and staples.

Attach cotter pin and screw eye to chain. Screw eye in bottom of desk.

Small Wardrobe

by Vivian Peritts

MATERIALS:
Mixed Nuts Tu-Door Armoire Cardboard Kraftable • 12" x 18" sheet of Brown kraft paper • Six 12" x 16" sheets of Natural mini-flute paper • One 12" x 16" sheet of Copper Corobuff paper • 8½" x 11" sheets of corrugated paper (4 sheets each Burgundy, Black and Ivory) • Pinking and scallop scissors • Rockies and Pompeii Canyon Cutters • Punchlines (¼" circle, ⅛" circle, teardrop) • Punches (heart, giant heart, jumbo heart) Tacky Glue • E6000 glue

INSTRUCTIONS:
Trace each piece before assembling armoire. Following manufacturer's instructions, assemble armoire. Keep all punch-out parts.
Doors - Refer to pattern on page 16 for colors and edging. Glue all pieces in place. Use 2 round punch-outs for handles and as décor on top. Glue all pieces with E6000.
Top and Bottom Shelves - Cover top of both pieces with Natural mini-flute corrugated paper. Edge with mini-flute corrugated paper. Cut as shown. Glue one giant Burgundy heart in center rounded section of shelves. Add a small Natural heart on top of the Burgundy heart. Leave back plain.
Top Décor - Using pattern on page 16, cut Top Décor piece from 2 Black 11" x 8½" corrugated sheets pieced together. Cut Brown kraft paper piece. Trim top edges of Black corrugated piece with pinking scissors and glue on front top. Trim top of kraft paper with Rockies scissors, punch a ¼" hole in each outward peak and glue over Black piece. Add trim to top edges as shown. Glue 3 medallions on top of both sides. Refer to pattern for colors, punches and scissors for triangular shape arrangements on bottom half of the Top Décor.
Sides - Cut strips of mini-flute the width of the edges and glue in place leaving slots open. Cover bottom legs on both sides with mini-flute. Cut two ½" x 11" strips of Burgundy corrugated paper. Wrap strips around center of legs. Refer to pattern on page 17 for remaining trim. Glue 3 medallions on each side piece.

A beautiful miniature chest... perfect to hold jewelry, photos and "what-nots"

MEDALLION PATTERN
FULL SIZE

Referring to photo, glue trims in place.

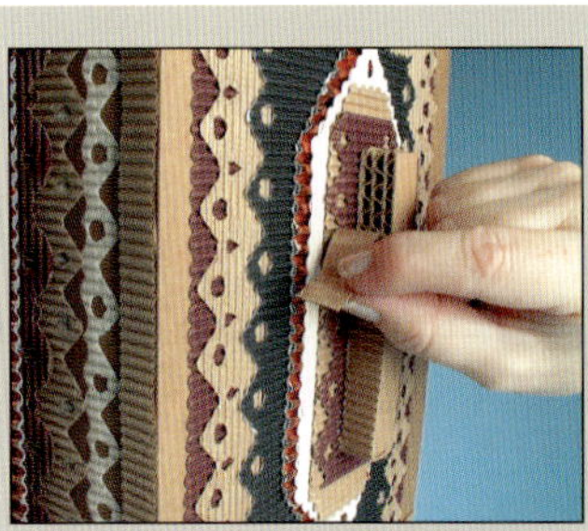

Glue paper strips to cover raw edges.

Stack and glue decorative medallions in place.

Center Line

Reverse for Other Half

Reverse for Other Half

Center Line

SIDE
PATTERN
(CUT 2)

Blue Chest

by Vivian Peritts

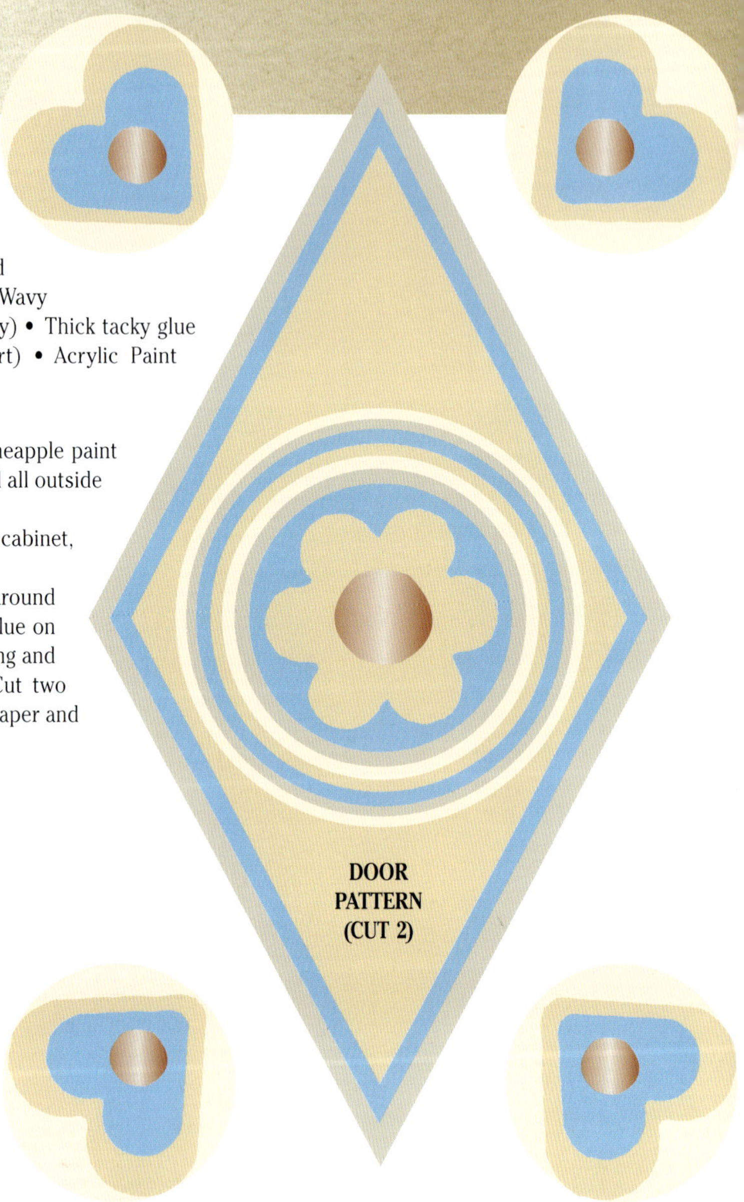

MATERIALS:
Walnut Hollow unfinished Tuckaway Cubby $10\frac{1}{4}$" x $10\frac{5}{8}$" x $4\frac{7}{8}$" • Unfinished wood button plugs (three $\frac{1}{2}$", eight $\frac{3}{8}$", eight $\frac{1}{4}$") • 12" x 16" sheet of Natural Wavy Ridges Paper • $8\frac{1}{2}$" x 11" sheets Corrugated Paper (2 Marigold, 2 Navy, 1 Ivory) • Thick tacky glue • Pinking scissors • Punches (giant flower, $1\frac{1}{4}$" circle, heart, jumbo heart) • Acrylic Paint (Pineapple, True Blue, White) • Medium sandpaper • Sponge

INSTRUCTIONS:
Prepare wood cabinet. Paint cabinet White inside and out. Let dry. Sponge Pineapple paint lightly over the inside, back and sides of cabinet. Paint doors inside and out and all outside parts of cabinet True Blue. Let dry. Lightly sand edges to give an aged look.

Edge Trim - Cut $\frac{1}{4}$" wide strips of Wavy Ridges paper and glue around the top of cabinet, down side fronts and across bottom below doors.

Diamond Front Trim - Punch 2 Marigold giant flowers, glue on Navy paper. Cut around Navy paper with pinking scissors to form a circle slightly larger than flower. Glue on Wavy Ridges paper and cut a circle slightly larger than Navy circle. Keep stacking and cutting the following layers: Ivory, Marigold, Navy, Natural, Ivory. Make 2. Cut two 3" x 5" Marigold diamonds. Trim with pinking scissors. Glue diamond on Navy paper and cut a second diamond slightly larger. Mount Navy diamond on Natural Wavy Ridges paper and cut last diamond. Make one of this combination for each door. Glue circles on top of diamonds and glue one on each door $\frac{3}{4}$" up from bottom.

Hearts & Circles - Make 8 ornaments to go around diamonds on door. Punch one Navy jumbo heart, glue on Marigold giant heart and glue on Ivory giant circle. Glue one on each side of bottom and top points of diamonds. Glue one $\frac{1}{2}$" button plug on top of Yellow flowers. Glue one $\frac{1}{4}$" button plug on top of each jumbo Navy heart.

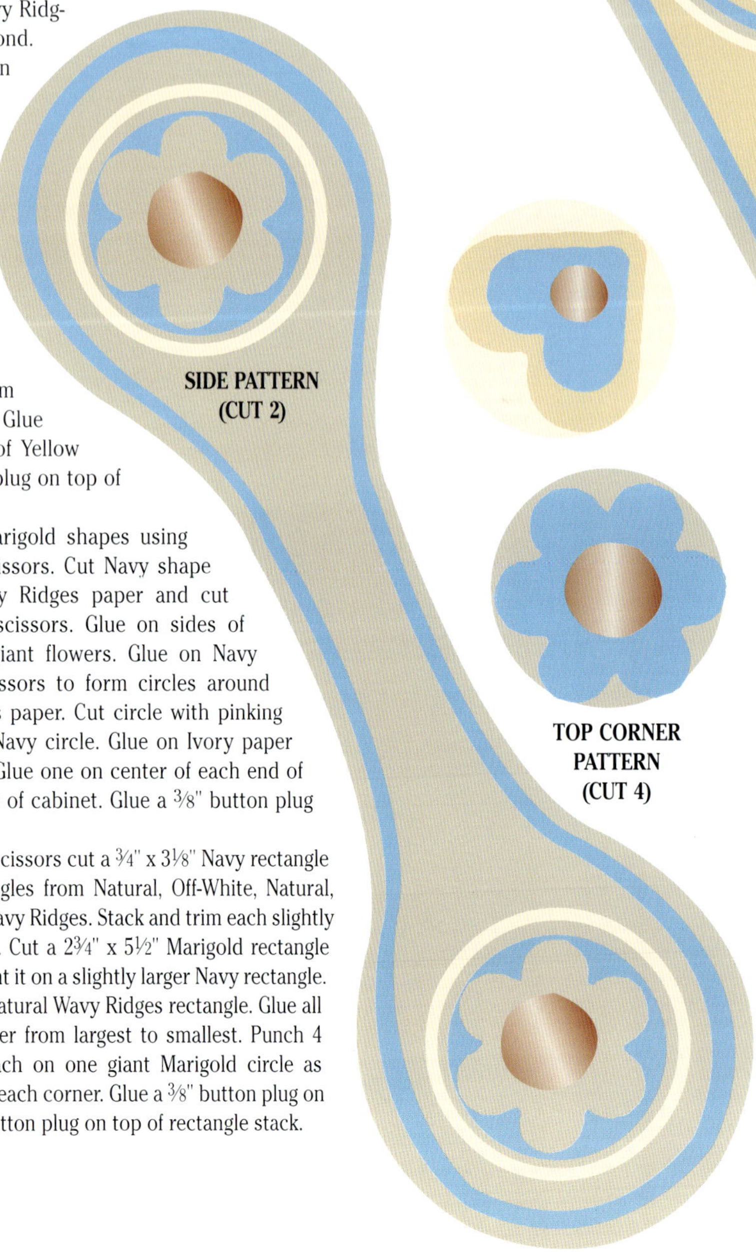

Sides of Cabinet - Cut 2 Marigold shapes using pattern, trim with pinking scissors. Cut Navy shape slightly larger. Glue on Wavy Ridges paper and cut slightly larger with pinking scissors. Glue on sides of cabinet. Punch 4 Marigold giant flowers. Glue on Navy paper. Trim with pinking scissors to form circles around flowers. Glue on Wavy Ridges paper. Cut circle with pinking scissors slightly larger than Navy circle. Glue on Ivory paper and cut again slightly larger. Glue one on center of each end of barbell shape design on sides of cabinet. Glue a $\frac{3}{8}$" button plug on top of each flower.

Top of Cabinet - With pinking scissors cut a $\frac{3}{4}$" x $3\frac{1}{8}$" Navy rectangle from Wavy Ridges. Cut rectangles from Natural, Off-White, Natural, Navy, Natural, and Off-White Wavy Ridges. Stack and trim each slightly larger than previous rectangle. Cut a $2\frac{3}{4}$" x $5\frac{1}{2}$" Marigold rectangle with pinking scissors and mount it on a slightly larger Navy rectangle. Mount these on a $3\frac{1}{2}$" x $6\frac{1}{2}$" Natural Wavy Ridges rectangle. Glue all rectangles on top of each other from largest to smallest. Punch 4 Navy giant flowers. Mount each on one giant Marigold circle as shown on pattern. Glue one to each corner. Glue a $\frac{3}{8}$" button plug on top of each flower and a $\frac{1}{2}$" button plug on top of rectangle stack.

TOP CORNER
PATTERN
(CUT 4)

Punch out circles and hearts for ornaments.

Stack and glue ornaments referring to photo.

by Vivian Peritts

Spool Dec-A-Lamp

MATERIALS:

Northland Spool Dec-A-Lamp • Natural colored kraft lampshade • 2 sheets of double-sided adhesive • One 12" x 16" sheet of Natural Wavy Ridges paper • 8½" x 11" sheet of Hunter Green Wavy Corrugated paper • 8½" x 11" sheet of Hunter Green Wavy Corrugated paper • Clear Gel Stain • Dark Green acrylic paint • Giant circle punch

INSTRUCTIONS:

1. Mix gel stain with Green acrylic paint. Paint wood portions of lamp. Wipe off the excess with a soft cloth. Let dry.
2. Cut Natural Wavy Ridges paper using paper pattern and apply to lamp.
3. Cut leaf sides A and B with corrugation pointing up on each side. Cut 7 sets of leaf halves. Apply to wavy Green paper with corrugation lines going up and down. Leave ⅛" space between halves. Trim around outside of leaves and apply to base of the lamp over Natural Wavy Ridges paper.
4. Punch approximately twenty 1¼" circles of Natural Wavy Ridges paper and glue on shade. Cut a few in half for bottom edge.

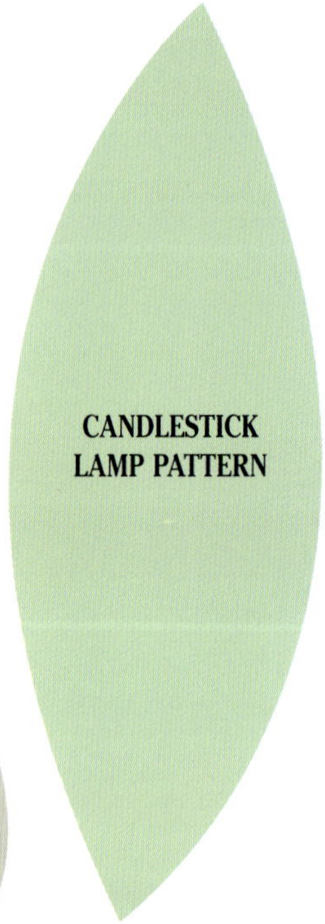

Lamp source:
Hollywood Lights
800 Wisconsin St. # D02104
Eau Claire, WI 54073

Candlestick Lamp and Shade

MATERIALS:

Candlestick lamp with wood base and Off White candle • White fabric lampshade • Two 12" x 16" sheets of Natural Wavy Ridges paper • Two 8½" x 11" sheets of Hunter Green Wavy Corrugated paper • Three 12" x16" sheets of double-sided adhesive • Emperors Gold Dazzling Metallics paint • Flat stiff paintbrush

INSTRUCTIONS:

1. Cut Natural Wavy Ridges paper to cover lamp as shown. For a lamp with different measurements, adjust pattern to fit.

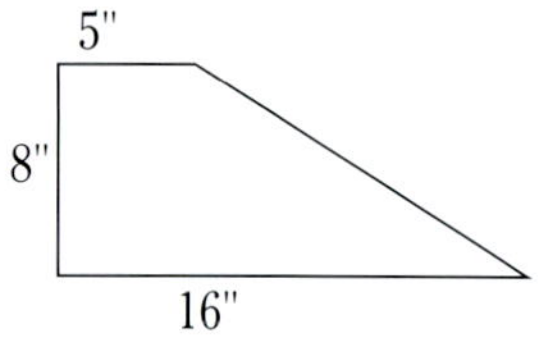

2. Dip brush in Gold paint and rub over a paper towel to remove excess. Drag paint brush over ridges and along top edge of paper.
3. Apply a 1" wide strip of adhesive over each side of the 8" and 7" ends of both pieces. Place 8" end on lamp and 7" end of second sheet on top and start wrapping both sheets together around lamp. Place a small piece of double-sided adhesive on the small end of both sheets to seal.
4. Back Green ridged paper with double-sided adhesive. Use Gold paint to accent ridges. Cut out 16 Green leaves with ridges going across leaves. Apply 4 in corners and 4 in center of each side. Apply 8 leaves above base over bottom of Natural wrap.
5. For lampshade, apply Gold paint to ridges of third sheet of Natural Wavy Ridges. Apply double-sided adhesive to back. Cut out 11 wavy shapes using pattern. Apply to lampshade with larger end at base of shade. Leave approximately 1¼" between strips.

Drag paintbrush over ridges and along top edge of paper.

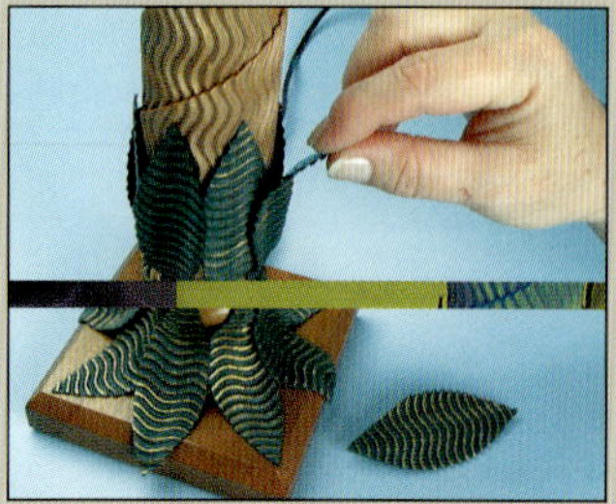

Cut leaves from painted paper. Glue in place.

16

Photo Album

by Vivian Peritts

MATERIALS:

10" x 12" Memory Album Cover with rippled edge • $8\frac{1}{2}$" x 11" filler paper • Four $\frac{3}{8}$" unfinished wood button plugs • $8\frac{1}{2}$" x 11" corrugated paper by (3 sheets each Burgundy and Natural kraft) • $8\frac{1}{2}$" x 11" Wavy Ridges Natural Kraft Corrugated paper • Pinking scissors and Rockies Canyon Cutters • Punches (giant heart, jumbo heart, craft heart) • Punchline $\frac{1}{4}$" circle • Tacky glue • Clear Gel Stain • Cadmium Yellow acrylic paint • Soft Rag

INSTRUCTIONS:

Prepare the wood. Mix equal parts of gel stain and paint. Paint wood. Wipe off excess with a soft rag. Let dry.

Left Front - Punch 2 giant Burgundy hearts and place one at each end of front. Glue with small end pointing in. Glue a jumbo Natural Wavy Ridges heart on top of each and add a small Burgundy heart. For center diamond shape, cut out each piece and trim with pinking scissors. Add a small Burgundy heart to top of diamond. Add a $\frac{1}{4}$" Burgundy circle diamond points.

Center - Trim an $8\frac{1}{2}$" x 11" sheet of Natural corrugated paper with Rockies scissors to $7\frac{3}{4}$" x $10\frac{3}{4}$". Punch a $\frac{1}{4}$" circle in each outward point of design. Glue to the front of the book. Refer to pattern for construction of remainder of design. Cut out patterns and trim with pinking scissors. Cut $\frac{1}{8}$" and $8\frac{1}{2}$" strips for curls, wrap around a pencil to shape. Glue wood button plugs on 4 sides of design.

Wrap paper strips around pencil to shape coils.

Keep your photos safe in this "look-like" carved wood album.

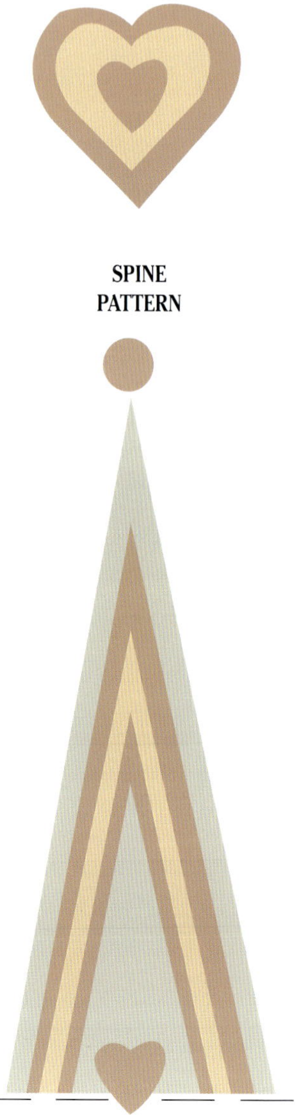

Small Albums

by Maria Filosa

MATERIALS:
Paper (ten assorted 9" x 12" sheets of Spectra Construction, 12" x 18" sheet of Brown or White Waffle, scrap pieces of Brown or White Mini Flute Corobuff, scraps pieces of Brown Wavy Ridges, 4-ply sheet of Spectra White Poster Board) • Mounting tape • Three 18" pieces of Brown raffia or one yard of 7/8" ribbon • One yard of Gold Cord for White/Gold album • Large Gold ribbon rose for White album • 4 small Gold Buttons with shanks removed for White album • 3 Brown buttons for Brown album • Embossing ink • Gold Embossing powder • Heat tool • Ruler • Scissors • 1/4" hole punch • Pencil or Pen • Hot glue

INSTRUCTIONS:
Covers - Cut two 5" x 8" pieces of poster board for top and bottom covers. On one piece, score a line 1" from 5" side, crease so that top cover of album opens easily.
Inside Pages - Cut sheet of construction paper in two 4¾" x 7¾" pieces.
Waffle Paper - Cut two 6" x 9" pieces of waffle paper. For White album, emboss tops of sheets with Gold embossing powder before assembling album. Apply mounting tape to all 4 sides of top cover, use back of fingernail to rub each strip onto paper. Remove protective backing from tape and place adhesive side down in center of a piece of Waffle Paper. Snip each corner of Waffle Paper at an angle as shown. Apply mounting tape to remaining side of top cover. Remove protective backing and fold over Waffle Paper pressing paper into tape. Apply mounting tape to all 4 sides of a piece of construction paper. Remove protective backing, center paper over inside of cover and press down. The construction paper will line the inside of the album covers. Repeat with bottom cover of album but do not score the edge.

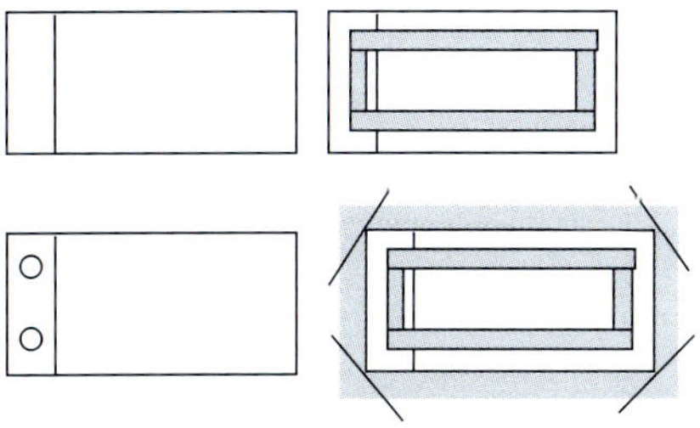

Holes - Referring to diagram, punch 2 holes in top cover along 1" scored section. Place bottom cover under top cover and mark holes with pencil. Punch bottom cover. Place one sheet of construction paper under top cover centering paper. Mark holes with pencil and punch through each mark. Place 4 sheets at a time under punched construction paper sheet. Punch through all sheets. Repeat with remaining construction paper sheets for 18 punched sheets.
Bind Album - Place construction paper sheets between both covers aligning all holes. Use raffia for Brown and ribbon for White album. Wrap ends of raffia together with clear cellophane tape. Lace raffia or ribbon from front of album through all holes to back of album and back through holes to front of album. Tie ends of raffia or ribbon in a knot then tie a bow. Trim excess raffia or ribbon.
Decorate Cover - Trim pieces of corrugated paper to desired shapes. Be creative! Place mounting tape or glue on the back of each piece and adhere to front cover.
Kite Shape - Use pattern to cut two pieces of Mini Flute Corobuff taking care to line up the 5¼" side with the ridges. Matching ridges, bring 2 longest sides together to form a kite. Glue on front cover. Glue buttons down center line.
Decorative Rolls - Cut three 3" x 4" strips of scrap Waffle paper. Roll strips over pencil and secure ends with mounting tape or glue. Slip roll off pencil, glue roll to front cover.
Heart - Cut cord into 3 equal pieces and braid together. Glue to front of White album in the shape of a heart. Glue a Gold ribbon rose in center of heart. Gold emboss ½" wide strips of White Mini Flute Corobuff and glue strips along edges of White album.

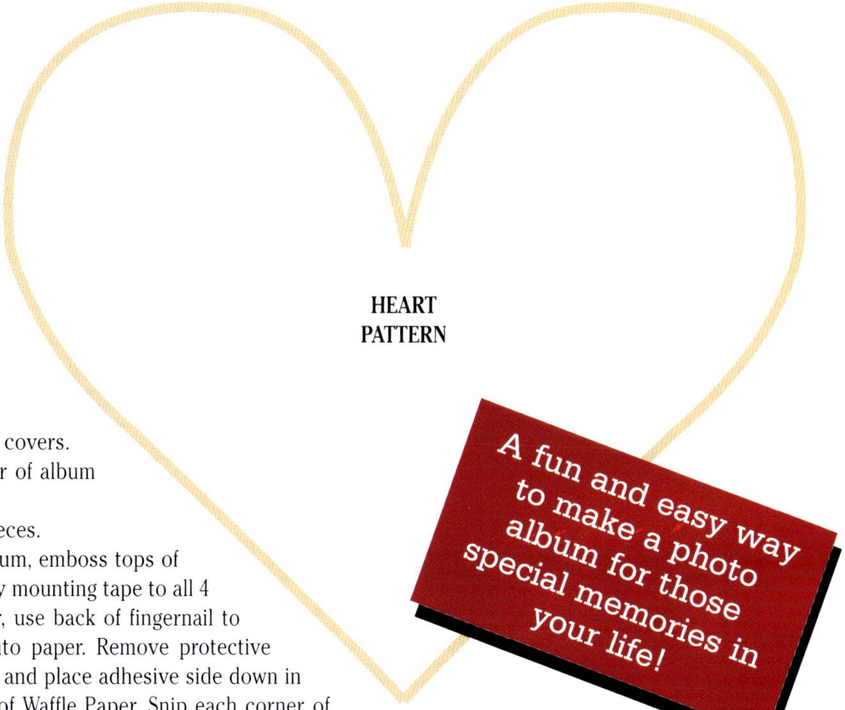

HEART PATTERN

A fun and easy way to make a photo album for those special memories in your life!

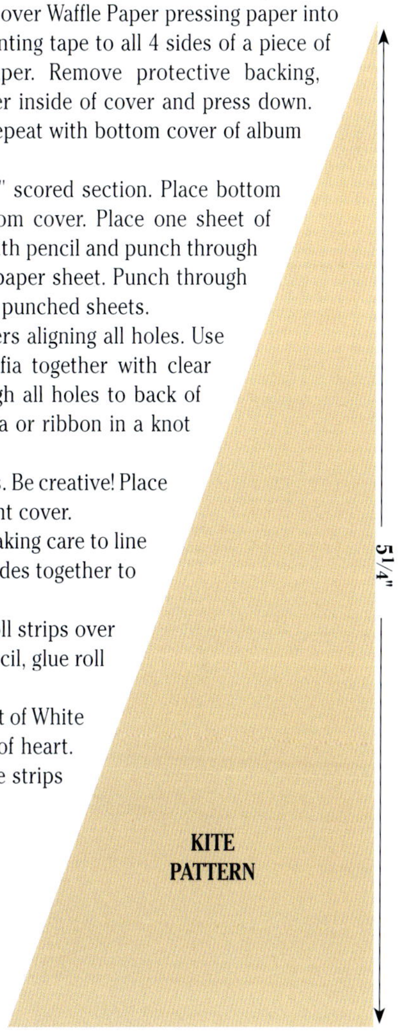

KITE PATTERN

Cover poster board covers with waffle paper.

Line inside of album cover with construction paper.

Punch holes in covers and pages. Bind and decorate album.

Portfolio Covers

by Maria Filosa

MATERIALS:
12" x 16" sheets of Corobuff or Mini Flute Corobuff • Cord, yarn or narrow ribbon for decorative closure • Large and small buttons • Hook and loop dots • Ink pads and embossing powders • Heat tool • Ruler • Scissors • Double sided tape or glue

INSTRUCTIONS:
1. Using ruler, measure and mark paper the width of the envelope by 3 times the length. For a 11" wide by 4" tall envelope, measure and cut paper to 11 " x 12".
2. Fold paper in thirds to finished size. Use double sided tape or glue to adhere inside edges. Cut sides of flap to form a point. Save the triangle pieces for another project or use to decorate front of envelope by bringing the ridges together to form a kite shape, glue in place.
3. For closure, glue a hook and loop dot under the flap or glue a button just below the flap. Wrap yarn, cord or string around envelope and button and tie in a bow.
4. To decorate, glue 3 buttons down the center of the kite shape. Or use a multi colored ink pad and rub color on ribs of the corrugated paper and emboss ribs using embossing ink and powders. Or glue a button on front and secure with elastic cord.

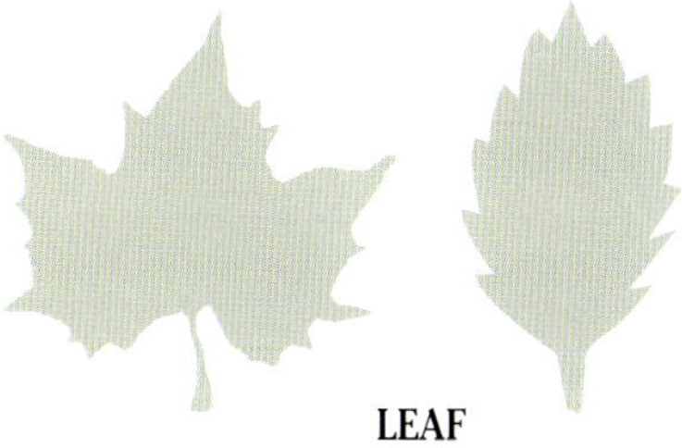

LEAF PATTERNS

These elegant and fun envelopes can be used to store special letters or treasures for yourself or the ones you love!

Cutting and folding diagram

Cut sides of envelope flap to form a point.

Matching ridges, form kite shape.

Glue kite shape on front of envelope.

Glue button in place and fasten with elastic cord.

Mini Box Necklace Pendants

by Maria Filosa

MATERIALS:
Mini Flute Corobuff or Wavy Ridges paper • Poster Paper • White glue • Scissors • Pen • Ruler • ³⁄₈" hole punch • Rattail cord • Beads • Hook and loop dots

INSTRUCTIONS:
1. Transfer pattern to poster paper and cut out pattern. Score all lines.
2. Place pattern on smooth side of paper sheets. Draw around outline of pattern and carefully fold pattern at each score line. Draw a heavy line with pen. This will score the paper enough to fold it on the line. Try not to punch through the paper. Remove pattern. Tip: Put a few pieces of double-sided tape on back of pattern and press tape against clothing to remove some of the tackiness. The pattern will stick to paper for tracing but will still be easy to remove.
3. Cut around outline with scissors. Punch 2 holes in each side as indicated on pattern. Fold all score lines.
4. Place glue on 2 long outside flaps marked A. Fold box so that both A flaps are glued to the edges marked B. Let dry.
5. Slip cord through holes in one side of box with a few inches of cord hanging from bottom hole. Tie overhand knot several inches from end of cord. Place 3 beads on cord and tie a second overhand knot to secure. Adjust length of cord as needed then repeat on the other side of box.
6. Place hook dot on inside of top flap and loop dot on front of the box so that hook and loop match when box is closed.
7. Decorate flap with painted wood hearts, lace, buttons, ribbon, charms, ribbon roses or stamped and embossed designs.
8. To decorate box, rub slow drying ink or embossing ink on ribs and sprinkle with embossing powder before assembly.

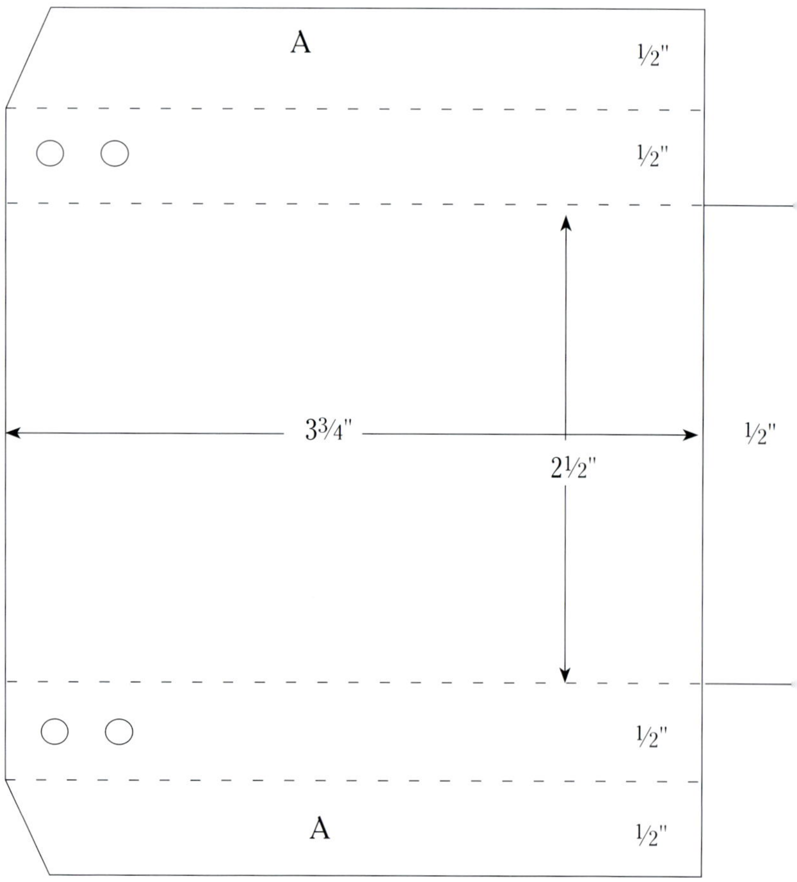

Use your Mini Box Pendant to carry business cards, credit cards or change for the phone. Kids will love them for carrying trading cards or small toys and trinkets!

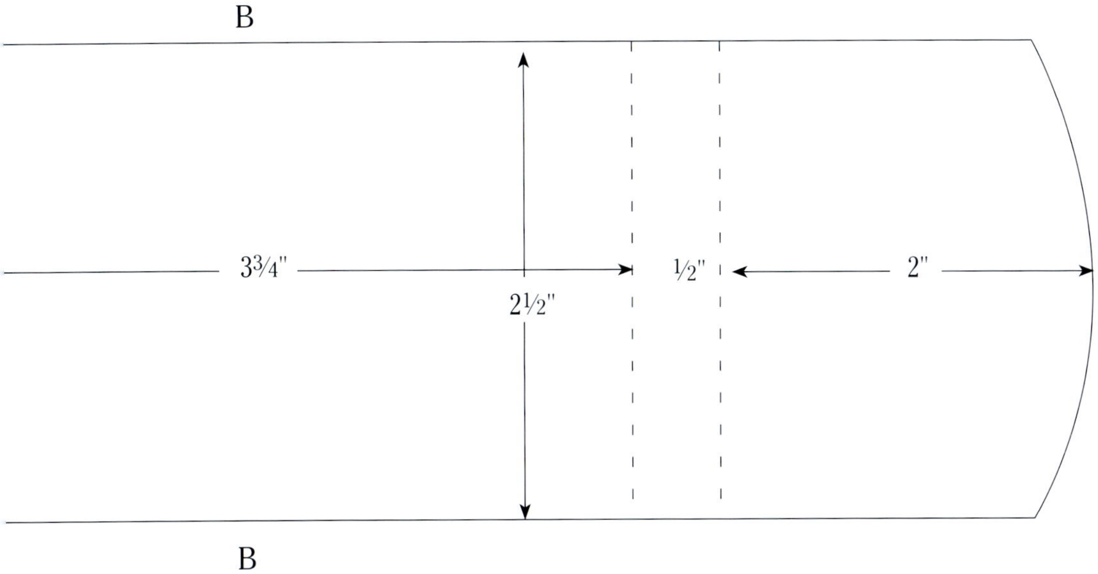

Place glue on flaps and assemble box.

Slip cords through holes in one side of box.

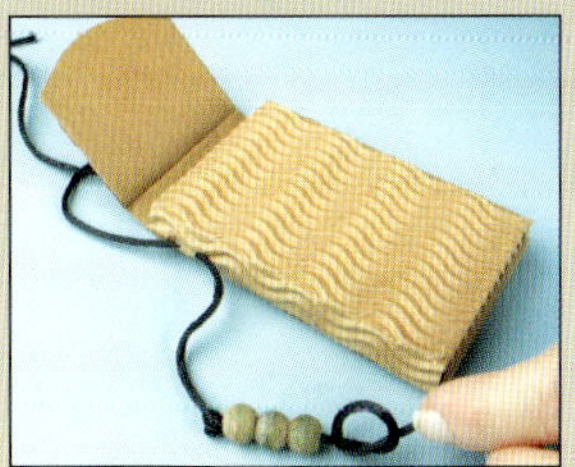
Tie overhand knot, string 3 beads and tie knot to secure.

Yellow Triangle Box

by Vivian Peritts

MATERIALS:

Unfinished 6½" x 4" triangle hinged box • Primary Yellow acrylic paint • Clear Gel Stain • Three 4 prong dinner forks and small 3 prong cocktail fork • 1½" gate hook and eye • 8½" x 11" sheets of corrugated paper (2 sheets each Marigold and Burgundy, 3 sheets Navy) • 12" x 16" sheet of Wavy Ridges Natural paper • Pinking scissors and Alps Canyon Cutters • Paintbrush • Tacky glue • Sturdy needle nose pliers • E6000 glue

INSTRUCTIONS:

Prepare wood. Mix equal parts of gel stain and Yellow. Use a paintbrush to apply the mixture to one section at a time. Wipe excess off with a soft cloth. Repeat until box has been painted inside and out. Let dry. Cut a sheet of Navy corrugated paper to match shape of box top. Trim with pinking scissors to make it ¼" smaller.

Cocktail Fork - Use a pair of needle nose pliers to curl outside fork tines back toward fork handle. Curl center tine slightly. Bend handle 1" up from fork wrist.

Top - Cut a Natural wavy ridge triangle the same size as the Navy triangle and trim with pinking scissors so that it is slightly smaller. Glue Navy triangle on top of box with tacky glue. Glue fork handle on top of Blue triangle with E6000 so that curled tines are on front point of box. Glue wavy Ridges triangle on top with tacky glue. Let dry. Place the latch on fork center tine and mark screw hole. Attach eye screw to box front. To finish top of box, cut 3 Burgundy shapes using pattern. Trim with pinking scissors. Glue Burgundy shapes to Wavy Ridges paper with tacky glue. Trim Wavy Ridges with pinking scissors slightly larger than Burgundy shape. Repeat gluing and cutting to form 3 identical shapes. Glue shapes on top of box making sure that each shape points to one of the corners and others meet in the center. The order of the stack is Burgundy, Natural, Navy, Marigold, Burgundy, Natural, Navy, Marigold and Burgundy.

Sides of Box - Do not start or end a strip of paper on a corner. It will require two 11" strips of paper to go all the way around the box.

Side of Lid - Cut bottom strip from Navy 1¼" wide after being trimmed with the Alps scissors. Glue around lid. The second layer should be Wavy Ridges ⅞" wide after being trimmed with pinking scissors. Cut third layer from Burgundy ½" wide after being trimmed with pinking scissors. Top layer is cut from Marigold and should measure ¼" cut with a straight edge.

Sides of Bottom Box - Cut four ⅞" Navy strips straight on one side and with Alps scissors on other side. Glue one strip ⅛" up from bottom edge of box and another one ⅛" down from top edge of box side. Place all strips with straight side of strip toward top and bottom edges of box. Cut four ⅝" Wavy Ridges strips with one straight edge and one pinked edge. Place one strip on top of each of the previous strips. Add ⅜" Burgundy strips cut with one straight and one pinked edge. Cut ¼" Marigold strip with 2 straight edges and glue it around center of box bottom.

Legs - Use needle nose pliers to bend outside tines of forks to front and 2 inside tines to back to make a platforms. Curl outside tines with pliers. Cut three 2" Navy strips and glue one around top of each fork handle. Cut three 1" Burgundy strips and glue one around each fork on top of Navy strips. Turn box upside down on a flat surface and glue legs to box with E6000. Let dry.

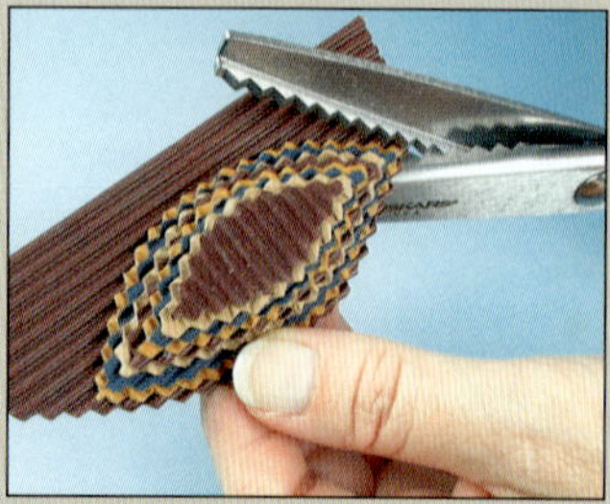

Make paper stacks cutting each layer slightly larger than the previous layer.

Stack and glue paper strips to decorate box.

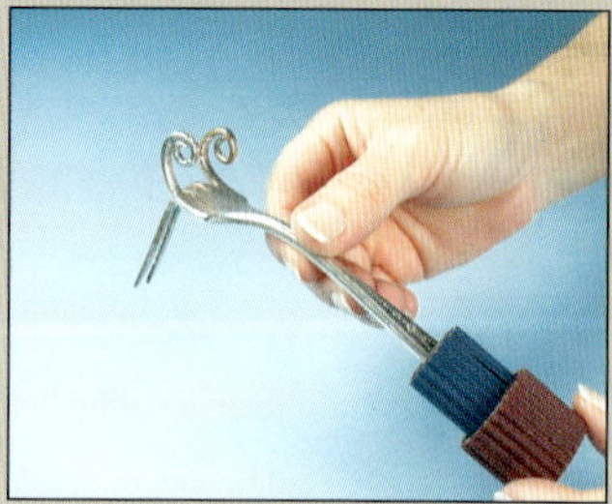

Bend fork tines. Wrap handle with paper strips to make leg.

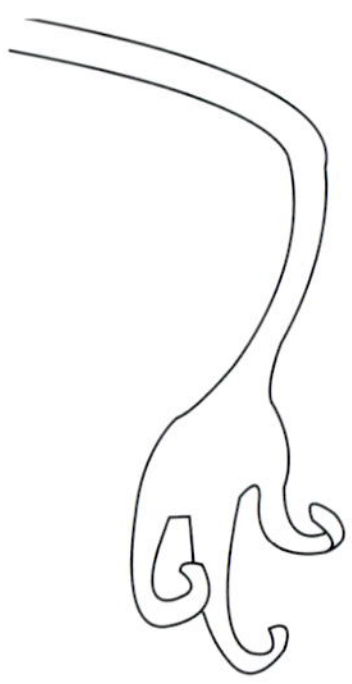

Corner fork detail

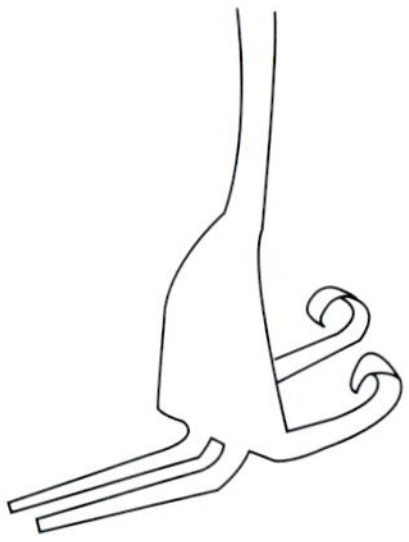

Leg detail

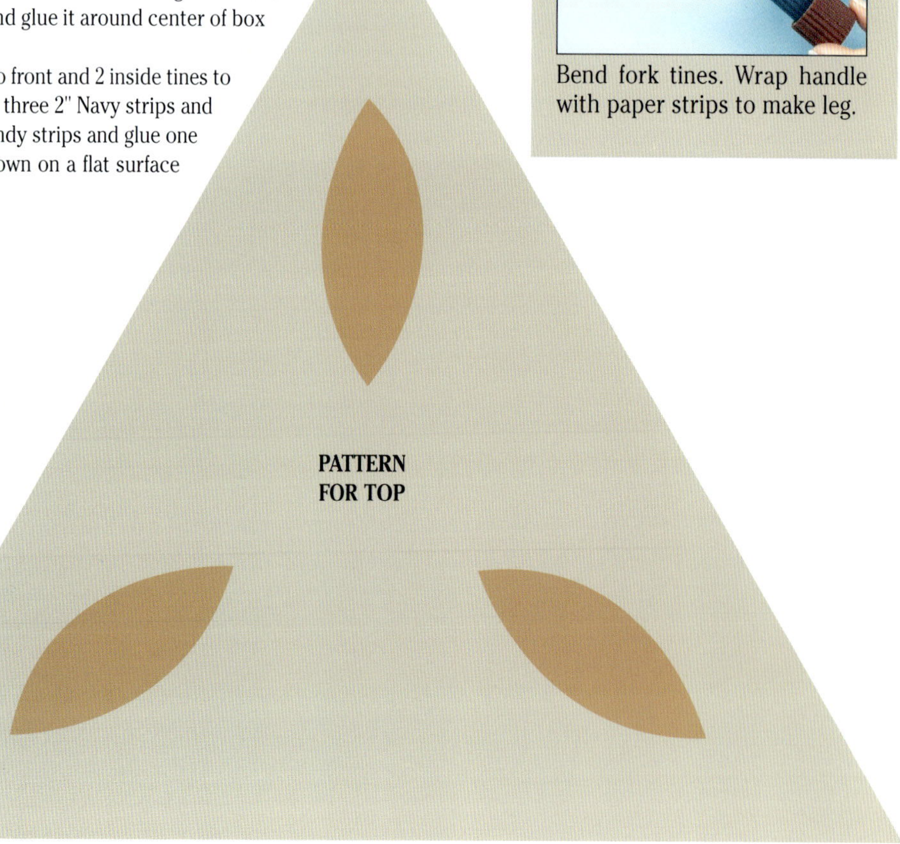

Candles

by Vivian Peritts

Candle Instructions

Pick candle in color desired or use candle painting medium and acrylic paint to change color.
Back all paper to be cut with PeelnStik.
Apply designs to the candle.
White wash candle using equal parts candle painting medium and White acrylic paint. Use a dry stiff brush and lightly stroke paint over the tops of the designs. Let dry.
Dip candle in the dipping wax. Dip each candle 3 times.

Techniques

- Use White wax for dipping wax. Blocks or slabs of wax may be purchased by the pound. Use a double boiler to melt wax. Be sure to pick a container to hold the wax that is taller and wider than the candles to be dipped. This container should fit into an even larger container that can be filled with water.
- To determine the amount of wax needed to dip candles, place the candle into the empty wax container and pour water into the container until it covers the candle. Remove the candle and mark the level of the water. Pour water out and dry the container. Place wax into the container and melt it. Add wax to reach marked level.
- Heat the wax to 190° on the candle thermometer. If melted wax appears cloudy it is not hot enough. If wax is steaming, it is too hot.
- Never leave wax unattended.
- To dip the candle, use pliers to hold the candle by the wick and lower the candle slowly into the melted wax. Let the candle dry between dips.
- To paint the candle, mix acrylic paint in the color desired with an equal amount of DecoArt Candle Painting Medium. Paint with a wide soft brush. Let paint dry before adding decorations.
- Apply PeelnStick double-sided adhesive to all papers being adhered to candles. PeelnStick is a strong adhesive and gives an even, secure bond between the candle and designs. When the candle is dipped in the wax, there will be no seepage of wax under the paper.

15" x 13" Peach 4-Sided Pyramid Candle

Use pinking scissors to cut wavy stems. Vary the length of flowers, stems and leaves. Cut all parts out of Wavy Ridges paper. Apply stems and leaves to all 4 sides of candle. Punch 4 giant hearts of Natural corrugated paper. Punch four 5⁄8" circles of Yellow corrugated paper. Center circles on front of flowers. Punch a 1⁄4" hole through center of all 4 flowers. Mount flowers on tops of stems. Punch 4 jumbo butterflies of Wavy Ridges paper and mount on each side of candle. Cut an 11" piece of Wavy Ridges paper. Use Alps decorative scissors to cut one long edge of strip. Apply around the base of the candle and cut off any excess.

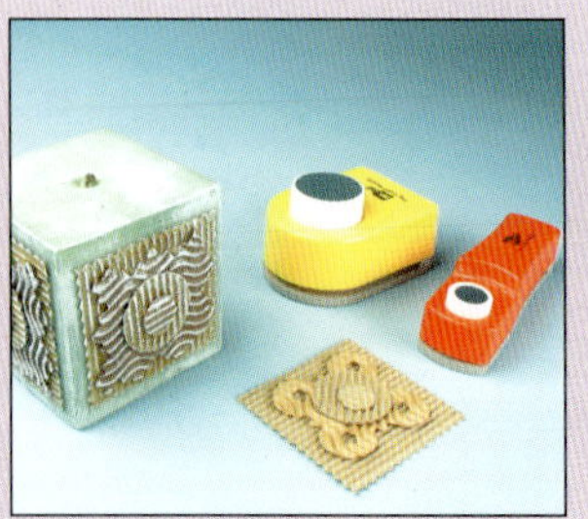

Cut and punch decorative shapes. Attach to candle.

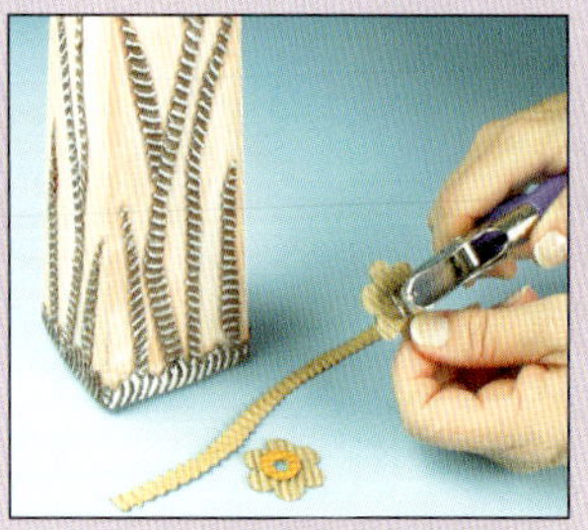

Cut leaves and punch flowers. Attach as shown.

To prevent accidents... never leave candles burning unattended when you leave the room.

3" x 6" Round Peach Candle

Bottom Layer - Cut a 4" Natural corrugated strip that goes around candle. Trim top and bottom of strip with Rockies scissors. Punch a square 1⁄4" hole in each peak. Apply to candle. Apply 1⁄4" circles in each downward peak of the design.
Second Layer - Cut a 2" wide strip. Cut top and bottom of strip using Pompeii scissors.
Third Layer - Punch 1 1⁄4" circles of Wavy Ridges paper and punch a 1⁄4" hole in center of each small circle. Apply to candle as shown.

3" x 3" Round Peach Candle

Cut all strips of Wavy Ridges paper.
Top Strip - Use Pompeii scissors and punch 1⁄4" holes.
Second Strip - Cut a 1⁄4" wide strip using pinking scissors.
Third Strip - Cut a 1" wide strip using Rockies scissors and punch 1⁄4" holes.
Fourth Strip - Cut strip using Pompeii scissors on top and straight scissors on bottom.

3" x 9" Round Green Candle

Cut strips of corrugated paper or Wavy Ridges using decorative edge scissors. Punch 1⁄4" holes or small stars in some strips. Apply strips diagonally to candle in both directions. Add hole punch-outs.

3" x 3" x 3" Green Candle

Bottom Layer - Cut a Natural corrugated square using pinking scissors.
Second Layer - Cut a Natural Wavy Ridges square with Alps scissors.
Third Layer - Punch 1 1⁄4" circles of Natural corrugated paper.
Top Layer - Punch 5⁄8" circle from Natural corrugated paper.
Back of Candle - Repeat design.
Sides - Turn all pieces in opposite directions. If the piece is straight up and down on the front and back, move it horizontally on the sides.

3" x 3" x 6" Peach Candle

Cut four 5" x 2" Natural corrugated rectangles with Alps scissors. Punch 1⁄4" holes in each point. Punch 3 giant flowers down centers of rectangles. Apply rectangles and punch-outs to candle. Punch 5⁄8" circles, apply in center of flowers.

3" x 3" x 6" Green Candle

Corners - Punch four jumbo hearts of Wavy Ridges paper.
Bottom Layer - Cut a rectangle of Natural corrugated paper using Pompeii scissors. Make the rectangle 4 large loops high and 2 loops wide. Punch a 1⁄4" hole in each large loop.
Second Layer - Cut a 6" x 1" strip of Wavy Ridges paper with Rockies scissors.
Top Layer - Punch one 1⁄4" circle on top of each of the sections of the second layer.
Outside Circles - Apply three 1⁄4" circle punch-outs on each side of design. Repeat on all 4 sides.

The open book will take on the shape of a star and can be displayed on a table or hung as an ornament.

Covers can be embellished with stamps, grapevine hearts, embossing ink, embossing powder, or use your own great ideas to make your Lotus Book special!